CONTENTS

INTRODUCTION	6
Chapter 1	7
Chapter 2	18

EDEDET ANDREW

SCIENTIFIC NAMES For Animals and Plants

SCIENTIFIC NAMES

FOR

ANIMALS AND PLANTS

Ededet Andrew

COPYRIGHT PAGE

All rights reserved under no circumstances shall any pert or full reproduction of this publication may permit without express permission of the author.

Copyright: 2024 Ededet Andrew

DEDICATION

This book is dedicated to the Spirit of the Lord for the inspiration and will power to put it together.

INTRODUCTION

The scientific names of the organisms follow immediately after the English or common names of such organism. Botanical names and zoological names are sometimes referred to as scientific names, scientific names are names given to organism in the scientific world by scientists. the names are mostly Greek, Latin or even the names of the scientist that discovered them.

The format of the names is that which has two names unlike that of humans with three names. The two name format is known as binomial nomenclature, the brainwork of the famous Carl Linneaus. Binomial refers to two while nomenclature could be substituted for the word names.

The first name of the scientific name is the generic of the organism. the first letter of the generic name is written in capital letter. the second name is the species name and has all its letters in lower case. this deals with the individual organism. Scientific names are always underlined, italicized or written in parenthesis.

CHAPTER 1

Scientific Names Of Some Selected Plants

1. Botanical Names For Some Selected Cereals

Cereals	Botanical Names
1. Guinea corn	*Sorghum* spp
2. Maize	*Zea mays*
3. Millet	*Pennisetum glaucum*
4. Hungry rice	*Digitaria exilis*
5. Wheat	*Triticum aestivum*
6. White rice	*Oryza sativa*
7. Brown rice	*Oryza glaberrima*

2. Botanical Name Of Some Selected Roots And Tubers

Plants	Botanical Names
1. Sweet potato	Ipomea batata
2. Carrot	Daucus carota
3. Irish potato	Solanum tuberosum
4. Sweet cassava	Manihot palmata
5. Bitter cassava	Manihot utilissima
6. White yam	Dioscorea rotundata
7. Water yam	Dioscorea alata
8. Trifoliate yam	Dioscorea domenterum
9. Aerial yam	Dioscorea bulbifera
10. Chinese yam	Dioscorea esculenta
11. Yellow yam	Dioscorea cayenensis

3. Botanical Names For Oil And Latex Crops

Plants	Botanical Names
1. Bambara nuts	Vigna subteranea
2. Avocado pea	Persea americana
3. Beniseed	Sesamum indicum
4. Castor	Ricinus cummunis
5. Coconut	Cocos nucifera
6. Melon	Citrullus vulgaria
7. Palm tree	Elaeis guinensis
8. Soya beans	Glycine max
9. Cashew	Anacardium occidentalis
10. Groundnut	Arachis hypogea
11. Rubber tree	Hevea brasilliensis

12. Sun flower   Helianthus annuus

4. Botanical Names For Some Selected Legumes

Legumes	Botanical Names
1. Soya beans	Glycine max
2. Cowpea	Vigna unguiculata
3. Brown beans	Vigna sinensis
4. Centro	Centrosema pubescens
5. Kudzu	Pueraria phaseoioides
6. Calapogonium	Calapogonium muconoides
7. Mucuna	Mucuna utilis
8. Sunhemp	Crotalaria juncea

5. Botanical Names For Some Selected Fruits

Fruits   **Botanical Names**

1. Mango — Mangifera indica
2. Sweet orange — Citrus sinensis
3. Sour orange — Citrus aurantium
4. Lime — Citrus aurantifolia
5. Lemon — Citrus limon
7. Tangerine — Citrus reticulata
8. Grapes — Citrus paradisi
9. Banana — Musa sapientum
10. Local pear — Dacryodes edulis
11. Paw-paw — Carica papaya
12. Guava — Psidium guajava
13. Pine-apple — Ananas comosus
14. Garden egg — Solanum melongena
15. Avocado pea — Persea americana
16. Cucumber — Cucumis sativum
17. African star apple — Chrysophyllum spp
18. Water melon — Citrullus lanatus
19. Date palm — Phoenix dactylifera
20. Coconut — Cocos nucifera
21. Bush mango — Irvingia gabonensis
22. Plantain — Musa paradisiaca
23. Baobab — *Adansonia digitata*
24. Melon — *Citrullus vulgaris*
25. Cashew fruit — *Anacardium occidentale*

SCIENTIFIC NAMES OF PLANTS AND ANIMALS

26. Shear butter fruit   *Vitellaria paradoxia*

6. Botanical Names For Some Selected Vegetables

Vegetables   Botanical Names

1. Tomato — Lycopersicum esculentum
2. Okra — Abelmoschus esculentum
3. Water-leaf — Talinum triangulare
4. Tridax — Tridax procumbens
5. Bitter-leaf — Vernona amygdalina
6. Cabbage — Brassica spp
7. Water lettuce — Pistia spp
8. Onions — Allium cepa
9. Roselle or sorrel — Hibiscus subdariffa
10. Crim-crim — Corchorius olitorius
11. Wild spinach[afang] — Gnetum africana
12. Spinach — Spinacia oleracea
13. Fluted pumpkin — Telfairia occidentalis
14. Silk cotton tree — Ceibas pentandra
15. Bombax plant — Bombax costatum
16. Atama leaf — Heinsia crinita
17. Utazi leaf — Gongronema latifolium
18. Editan leaf — Lasianthera africana
19. Oha leaf — Pterocarpus mildbraedii

7. Botanical Names For Some Selected Sugar Crops

Sugar Crop	Botanical Names
1. Sugar-cane	Saccharum officinarum
2. Sugar-beet	Beta vulgaris

8. Botanical Names For Some Selected Spices

Spices	Botanical Names
1. Sweet pepper	Capsicum annum
2. Ginger	Zingiber officinale
3. Lemon grass	Cymbopogon spp
4. Locust beans	Parkia spp
5. Mint	Mentha spp
6. Garlic	Allium sativum
7. Onions	Allium cepa
8. Green pea	Pisum sativum
9. Curry	Murraya koenigii
10. Basil (Scent leaf)	Occimum basilicum
11. Iron tree	Prosopis africana

9. Botanical Names For Some Selected Forage Crops

Grasses	Botanical Names
1. Elephant grass	Pennisetum purpureum
2. Guinea grass	Panicum maximum
3. Gaint star grass	Cynodon plectostachyus
4. Bahama grass	Cynodon dactylon
5. Carpet grass	Axonopus compressus
6. Spear grass	Imperata cylindrica
7. Northern gamba grass	Andropogon gayanus
8. Southern gamba grass	Andropogon tecorum
9. Rhodes grass	Chloris gayana
10. Pig weed	Boerhavia diffusa
11. Wild marigold	Aspilia africana
12. Crow foot grass	Eleusine indica
13. Tridax	Tidax procumbens
14. Water grass	Commenlina vogelli
15. Broom weed,	Sida acuta
16. Stiga/witch grass	Striga spp
17. Pig weed	Amaranthus spinosus
18. Goat weed	Ageratum conyzoides
19. Siam weed	Eupatorium odoratum
20. Kudzu	Pueraria phaseoloides
21. Milk weed	Asclepias syriaca

10. Botanical Names For Some Selected Fibres

Fibres	Botanical Names
1. Cotton	*Gossypium sativum*
2. Kenaf	*Hibiscus cannabicus*
3. Sisal	*Agave sisalana*
4. Jute/crim-crim	*Corchorus olitorius*
5. African fan palm	*Borassus aethiopum*

11. Botanical Names For Some Selected Ornamental

Ornamental	Botanical Names
1. Rose flower	*Rosa* spp
2. Hibscus flower	*Helianthus aunnas*
3. Pride of barbados	*Caesalpinia pulcherrima*
4. Sun flower	*Halianthus annuus*
5. Wild marigold	*Aspilia africana*
6. Ixora	*Ixora coccinea*

12. Botanical Names Of Some Selected Beverages And Stimulants

Plant   **Botanical Names**

1. Cocoa — Theobroma cacao
2. Tea — Camellia sinensis
3. Coffee — Coffee arabica
4. Kola — Cola nitida
5. Tobacco — Nicotiana tobaccum
6. Marijuana — Cannabis sativum

13. Botanical Names For Some Seleted Medicinal Crop

Plants   **Botanical Names**

1. Neem plant — Azadirachta indica
2. Moringa — Moringa oleifera
3. Bitter cola — Garcinia cola
4. Lemon grass — Cymbopogon spp

CHAPTER 2

Scientific Names Of Some Selected Animals

1. Scientific Names For Some Selected Farm Animals

Animals   **Scientific Names**

1. Snail — Achatina achatina
2. Bee — Apis mellifera
3. Cow — Bos spp
4. Pig — Sus domesticus
5. Sheep — Ovis airies
6. Goat — Capra hircus
7. Domestic fowl — Gallus domesticus
8. Domestic duck — Anas platyrhynchos
9. Crayfish — Orconectes obscurus
10. Cat-fish — Clarias gariepinus
11. Crab — Cancor spp
12. Guinea fowl — Numice melegris
13. Saw-fish — Pristia spp
14. Scaleless dragon-fish — Chirostomias pliopterus

2. Scientific Names For Some Selected Pest And Insects

Animals	Scientific Names
1. Black-fly	Simulum damnosum
2. Cockroach	Periplaneta americana
3. Beans weevil	Callosobrucus maculatus
4. Tse-tse fly	Glossina spp
5. Cricket	Acheta domesticus
6. Cotton stainer	Dysdercus spp
7. Aphids	Aphis spp
8. Locust	Schistocerca gregaria
9. Maize weevil	Sitophilus zeamais
10. Rice weevil	Sitophilus oryzae
11. Termite	Macrotermes bellicosus
12. Grasshoper	Zonocerus variagatus
13. Fruit-fly	Drosophila melanogaster
14. Stem borer	Sesamia calamistis
15. White-fly	Bemisia tabacci
16. Yam beetle	Heteroligus meles
17. Tick	Boophilus decoloratus
18. Liverfluke	Fasciola hepatica
19. Cocoa meadlybug	Planocecus citri
20. Leat beetle	Crioceris livida
21. Rhinoceros beetle	Oryotes rhinoceros
22. Capsid(kola nut)	Sahibergella singularis

23. Onions thrips	*Thrips tabaci*
24. Water-flea	*Daphnia pulex*
25. Water scorpion	*Nepa spp*
26. Sweet potato weevil	*Cylas brumeus*

3. Causative Agents And Diseases Of Some Selected Major Crops

Disease	Causative Agent
1. Bacterial blight of cassava	Xanthomonas manihot
2. Bacterial blight of cotton	Xanthomonas malvacearum
3. Maize smut	Ustilago maydis
4. Rust disease of maize	Puccimia sorghi
5. Rice leaf spot	Cercospora oryzae
6. Sorghum head smut	Sphacetatheca reliana
7. Groundnut tikka disease	Cercospora persorata
8. Groundnut asflo-toxin	Aspergillus flavus
9. Cocoa blackpod	Phytopthora palmivora
10. Banana/plantain panama disease	Fusarium oxysporum
11. Banana/plantain sigatoka disease	Mycosphacrella musicola
12. Gummosis disease of citrus	Phytophthora citrophthora
13. Bacterial wilt of tomato	Bacillus bacteria
14. Groundnut rosette	Groundnut rosette virus
15. Tomato root knot	Nematodes

4. Scientific Names For Some Selected Wild And Other Animals

Animals	Scientific Names
1. African elephant	Loxodonta africana
2. Buffalo	Syncerus spp
3. Crocodile	Crocodylus nitoticus
4. Cattle egret	Bubuluis ibis
5. Chimpanzee	Pan troglodytes
6. Common chameleon	Chameleo chameleo
7. Dog	Cannis domesticus
8. Gorilla	Gorilla gorilla
9. House centipede	Cermacha forceps
10. Horse	Equus ungulata
11. Lion	Panthera leo
12. Leopard	Panthera pardus
13. Tiger	Panthera tigris
14. Millipede	Marceus spp
15. Parrot	Psittacus errithacus
16. Toad	Bufo bufon
17. Frog	Rana temporaria
18. Tape worm	Taenia solium
19. Earthworm	Ascaris lumbricoides
20. Spotted hyena	Crocuta crocuta
21. Giraffe	Giraffa camelopardali
22. Wild pig	Sus scrofa
23. Bush fowl	Francolinus bicalcarcotus

24. Ostrich	Struthio camelus
25. Kiwi	Apteryx spp
26. Black vulture	Coragyps atratus
27. Quelea bird	Quelea quelea
28. Weaver bird	*Ploceus cucullatus*
29. Hawk	*Buteo spp*
30. Scorpion	Pterois radiata
31. Lizard	Agama agama
32. Domestic cat	*Felis catus*
33. Barren owl	*Strix varia*
34. Turkey	*Meleagris gallopavo*
35. Bush rat	*Rattus rattus*
36. Zebra	*Equus simplicidens*
37. Spider	*Nephilla clavipes*